My Book

This book belongs to

Name: _____

Copy right © 2019 MATH-KNOTS LLC

All rights reserved, no part of this publication may be reproduced, stored in any system or transmitted in any form, or by any means, electronic, mechanical, photocopying, recording, or otherwise without the written permission of MATH-KNOTS LLC.

Cover Design by :
Gowri Vemuri

First Edition :
January , 2020

Author :
Gowri Vemuri

Edited by :
Raksha Pothapragada
Ritvik Pothapragada

Questions: mathknots.help@gmail.com

NOTE : VDOE is neither affiliated nor sponsors or endorses this product.

Dedication

This book is dedicated to:

My Mom, who is my best critic, guide and supporter.

To what I am today, and what I am going to become tomorrow, is all because of your blessings, unconditional affection and support.

This book is dedicated to the
strongest women of my life,
my dearest mom
and
to all those moms in this universe.

G.V.

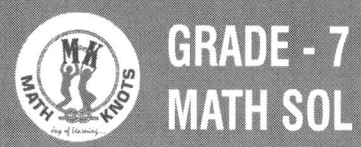

GRADE - 7 MATH SOL

Test Details

The Virginia Board of Education and Virginia Department of Education (VDOE) have developed the Virginia Assessment Program (VAP) to measure and evaluate students' academic progress in the Standards of Learning (SOLs). The SOLs indicate Virginia's expectations for what students should know and be able to do in the subject areas of reading, writing, mathematics, science, and history/social science.

Students in grades 3-12 to take the Standards of Learning (SOL) assessments each year. Some of the tests are required by all students each year, and others are required only at specific grade levels. Additionally, with the removal of some SOL tests in recent years, the VDOE assigned the responsibility of the creation and administration of alternate, performance - based assessments on local divisions. Student scores from these tests determine a school's and the division's state accreditation and measures progress toward meeting federal targets.

Virginia Standards of Learning (SOL) tests are generally given online unless a student has an identified and documented need to be assessed using paper, pencil format. The test question format is typically multiple choice, and each test contains some technology enhanced items.

GRADE 3-8 STANDARDS OF LEARNING(SOL) TESTS

GRADE 3	GRADE 4	GRADE 5	GRADE 6	GRADE 7	GRADE 8
	VIRGINIA STUDIES				WRITING
MATH	MATH	MATH	MATH	MATH	MATH
READING	READING	READING	READING	READING	READING
		SCIENCE			SCIENCE

NOTE : VDOE is neither affiliated nor sponsors or endorses this product.

GRADE - 7 MATH SOL

Test Details

END OF COURSE STANDARDS OF LEARNING (SOL) TESTS

GRADE 9	GRADE 10	GRADE 11
ALGEBRA I	GEOMETRY	ALGEBRA II
EARTH SCIENCE	BIOLOGY	CHEMISTRY
WORLD HISTORY I	WORLD HISTORY II	VIRGINIA & US HSTORY
		WORLD GEOGRAPHY
		ENGLISH : READING
		ENGLISH : WRITING

Any Student taking one of the courses listed here is expected to take the corresponding end-of-course SOL test. The grade levels depicted here represent grade level at which students typically participate in these courses.

NOTE : VDOE is neither affiliated nor sponsors or endorses this product.

GRADE - 7 MATH SOL

Test Details

SOL Test Scoring and Performance Reports:

Standards of Learning assessments in English reading, mathematics, science and history/social science are made up of 35-50 items or questions that measure content knowledge, scientific and mathematical processes, reasoning and critical thinking skills. English writing skills are measured with a two-part assessment that includes multiple-choice items and an essay.

Student performance is graded on a scale of 0-600 with 400 representing the minimum level of acceptable proficiency and 500 representing advanced proficiency. On English reading and mathematics tests, the Board of Education has defined three levels of student achievement: basic, proficient, and advanced, with basic describing progress towards proficiency.

Performance Achievement Levels:

- The achievement levels for grades 3-8 reading and mathematics tests are: *Pass/Advanced, Pass/Proficient, Fail/Basic,* and *Fail/Below Basic*.

- The achievement levels for science tests, history tests, and End-of-Course (EOC) tests* are: *Pass/Advanced, Pass/Proficient,* and *Fail/Does Not Meet*.

- The EOC Writing (2010 SOL) test, EOC Reading (2010 SOL) test, and EOC Algebra II (2009 SOL) test have an achievement level of *Advanced/College Path* in place of the *Pass/Advanced* achievement level.

NOTE : VDOE is neither affiliated nor sponsors or endorses this product.

GRADE - 7 MATH SOL

INDEX

Contents	Page No
Preface	1 - 12
Formulae	13 - 16
Practice Test - 1	17 - 34
Practice Test - 2	35 - 52
Practice Test - 3	53 - 70
Practice Test - 1 Answer keys	71 - 76
Practice Test - 2 Answer keys	77 - 80
Practice Test - 3 Answer keys	81 - 84
Score calculation	85 - 90

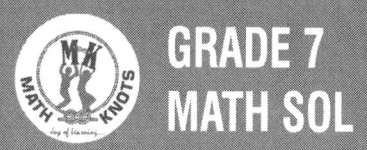

GRADE 7 MATH SOL

FORMULAE

FORMULA SHEET

1. Area of a triangle

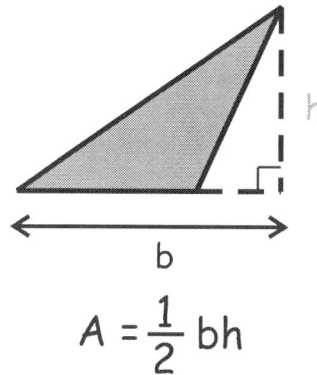

$A = \frac{1}{2} bh$

2. Area of a parellelogram

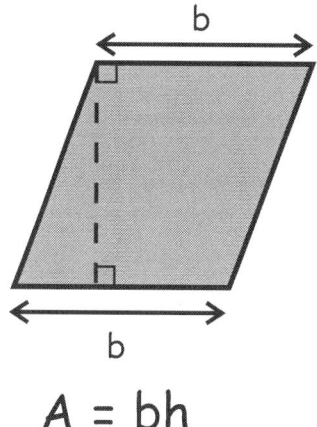

$A = bh$

3. Volume and Surface area of a cuboid

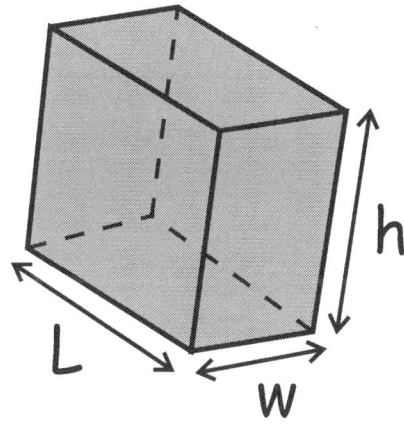

$V = lwh$
$S.A = 2(lw + lh + wh)$

4. Perimeter and Area of a Square

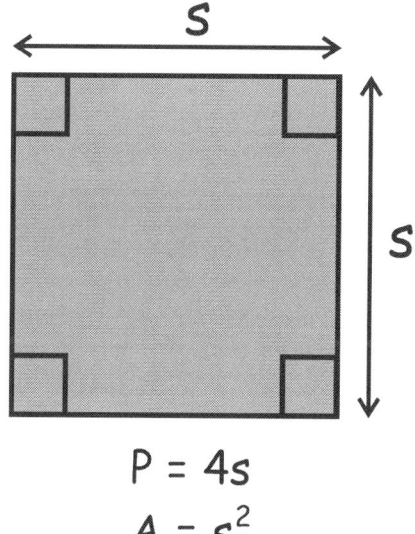

$P = 4s$
$A = s^2$

GRADE 7 MATH SOL

FORMULAE

5. Area of a Trapezium

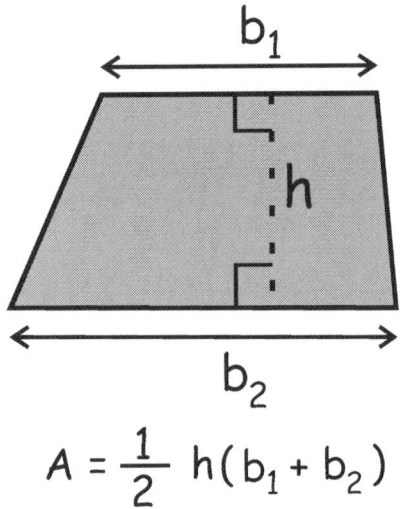

$$A = \frac{1}{2} h(b_1 + b_2)$$

6. Circumference and Area of a Circle

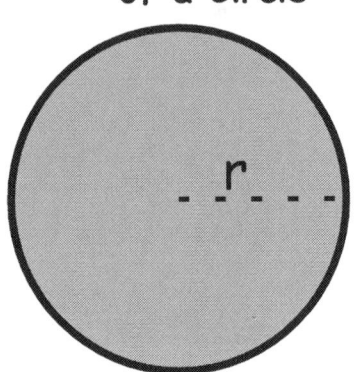

$c = 2\pi r$
$A = \pi r^2$

pi

$\pi = 3.14$
$\pi = \frac{22}{7}$

7. Perimeter and Area of a Rectangle

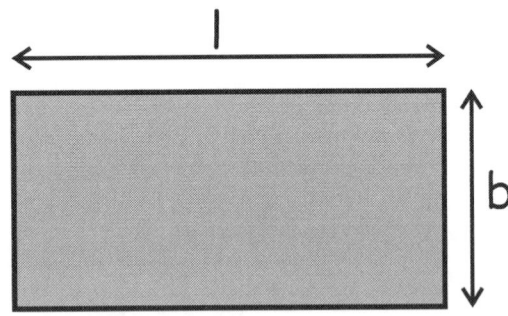

Area = l b   ×
Perimeter = 2(l + b)

8. Acute angle

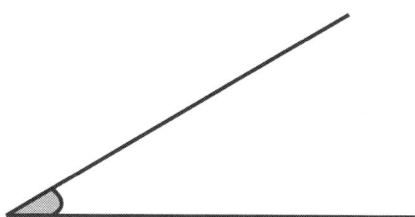

Any angle less than 90⁰
is called as acute angled

9. Right angle

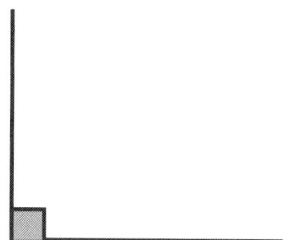

An angle equal to 90⁰
is called as right angled

10. Obtuse angle

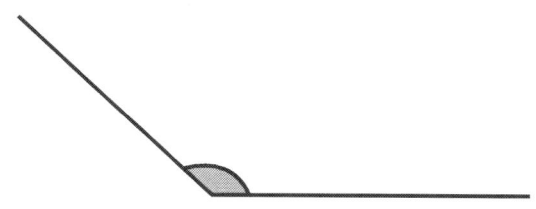

Any angle greater than 90⁰
is called as obtuse angled

11. Perimeter of a polygon KLMNOP

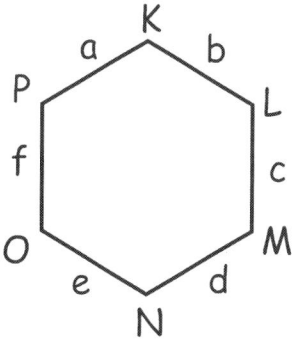

Perimeter =
 (a + b + c + d + e + f) units

GRADE 7 MATH SOL

Acronym

Abbreviations

milligram	mg	volume	V
gram	g	total Square Area	S.A
kilogram	kg	area of base	B
milliliter	mL	ounce	oz
liter	L	pound	lb
kiloliter	kL	quart	qt
millimeter	mm	gallon	gal.
centimeter	cm	inches	in.
meter	m	foot	ft
kilometer	km	yard	yd
square centimeter	cm^2	mile	mi.
cubic centimeter	cm^3	square inch	sq in.
		square foot	sq ft
		cubic inch	cu in.
		cubic foot	cu ft

year	yr
month	mon
hour	hr
minute	min
second	sec

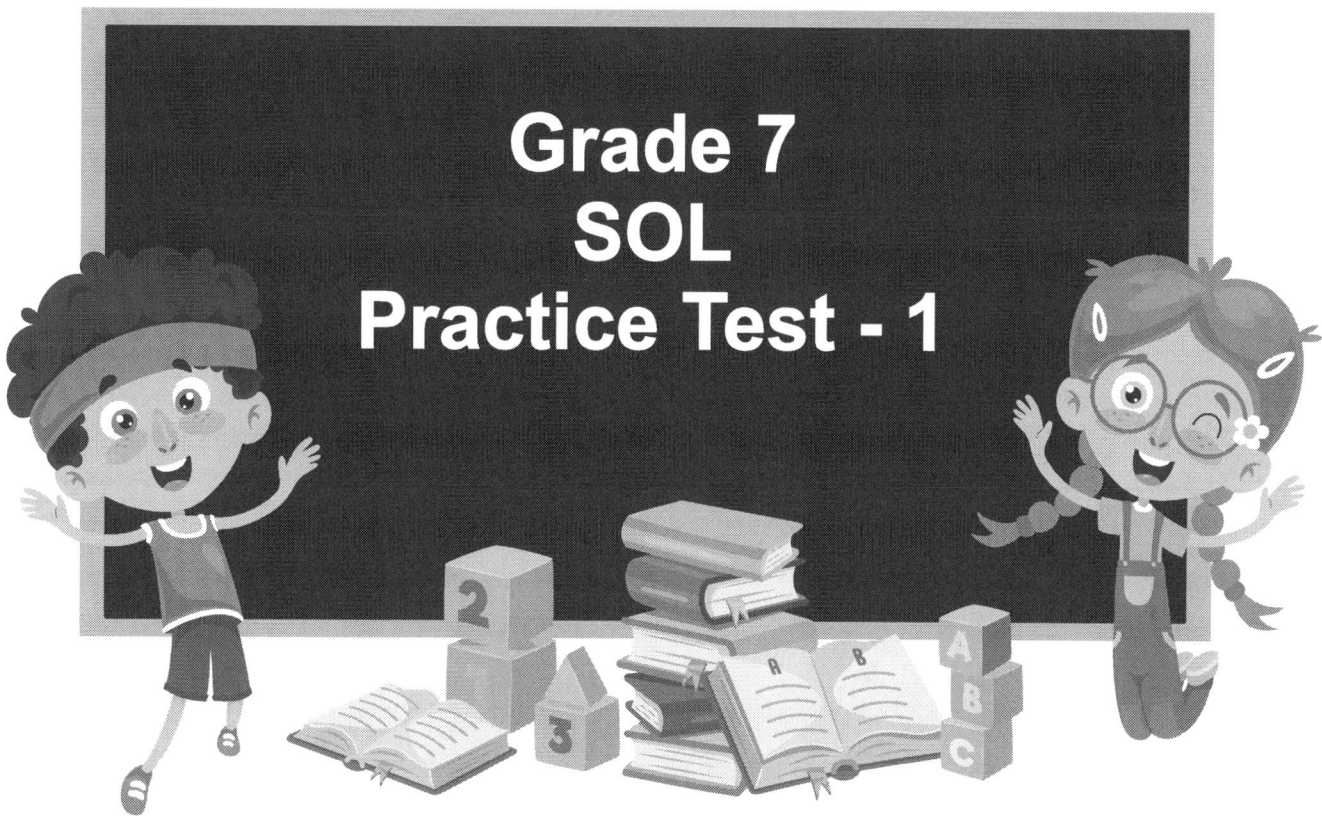

Grade 7 SOL

Vol 1 Test 1

1. Simplify the below

$$\left| \frac{((-43) - (-1))}{7} \right|$$

   (A) -7 (B) 19

   (C) 6 (D) 20

2. Simplify (-10) - (-15)

   (A) 2 (B) 5

   (C) -4 (D) 13

3. Express the below number in scientific notation

   990000

   (A) 9.9×10^{-5} (B) 9.9×10^{-6}

   (C) 9.9×10^{5} (D) 9.9×10^{6}

4. A rectangle is 25 in tall and 35 in wide. If it is reduced to a height of 5 in, then how wide will it be?

   (A) 25 in (B) 175 in

   (C) 10 in (D) 7 in

5. Simplify the below

   $((-40) \times 2) \div 20$

   (A) -4 (B) 15

   (C) -3 (D) 10

6. Which of the below is an example of "Rotation 180° about the origin"?

(A)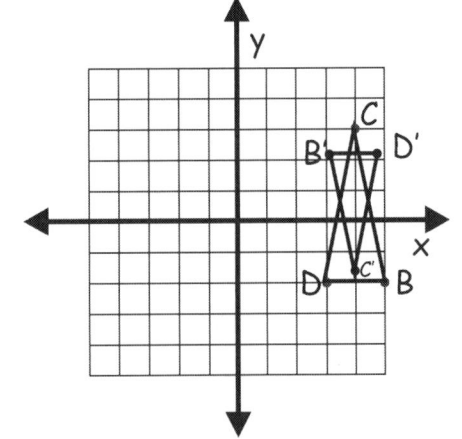

(B)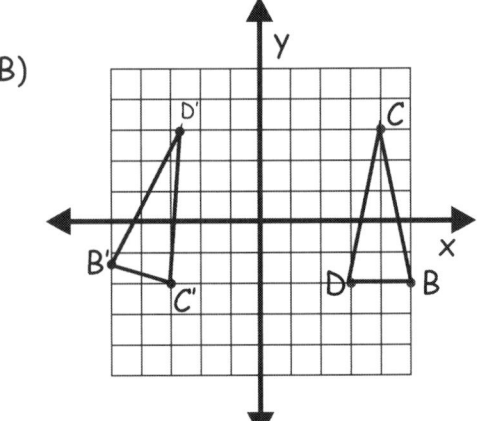

(C)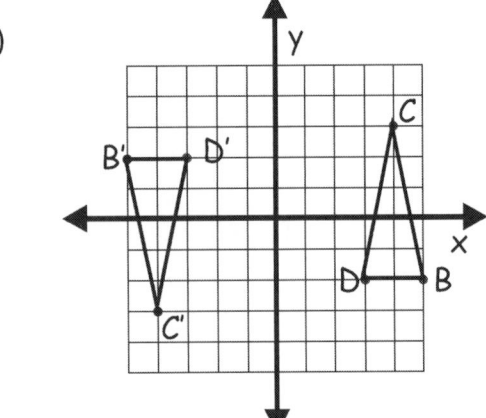

(D)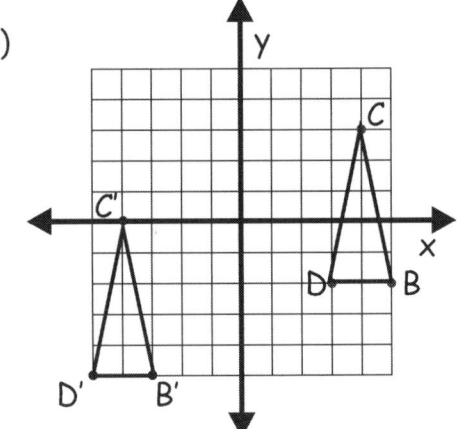

7. Determine the quadrant to which the below point belongs to?

(5, 1)

(A) Quadrant I

(B) Quadrant II

(C) Quadrant III

(D) Quadrant IV

Grade 7 SOL

Vol 1 Test 1

8. Find the volume of the cylinder with a radius of 5 cm and a height of 5 cm.

   (A) 329.9 cm^3 (B) 280.4 cm^3

   (C) 140.2 cm^3 (D) 392.7 cm^3

9. Find the area of the enclosure.

   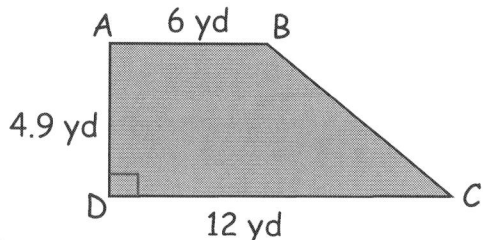

   (A) 50.2 yd^2 (B) 22.1 yd^2

   (C) 88.2 yd^2 (D) 44.1 yd^2

10. Find the total surface area of the given three dimensional shape.

    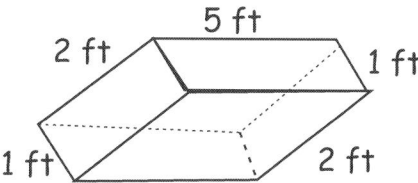

    (A) 71 ft^2 (B) 34 ft^2

    (C) 35.5 ft^2 (D) 39.4 ft^2

11. In how many ways the following task can be done ?
    A team of 6 basketball players need to choose a captain and co-captain.

    (A) 30 (B) 36

    (C) 32 (D) 26

12. Which of the below scatter plot can be plotted using the given data ?

x	30	40	44	44	50	55	60	64	87	100
y	65	57	54	61	55	45	43	45	21	13

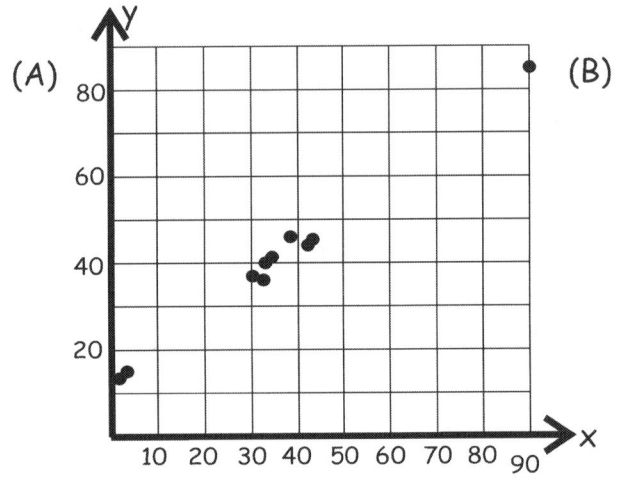

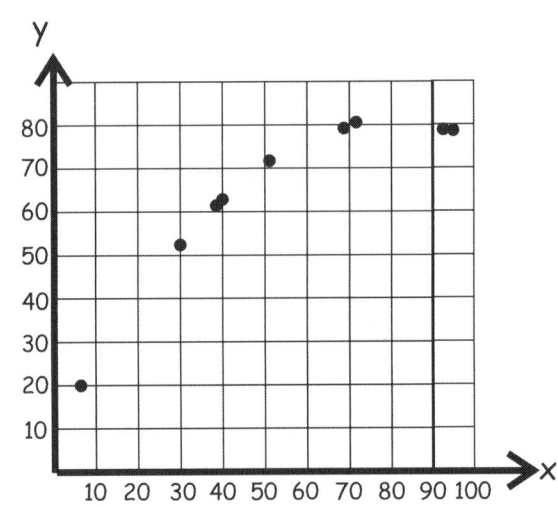

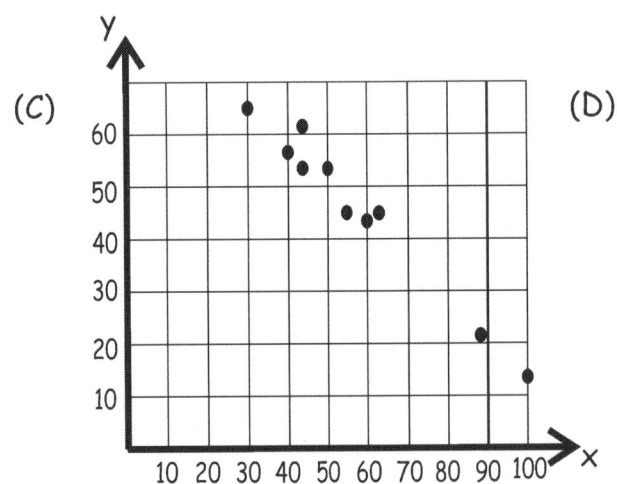

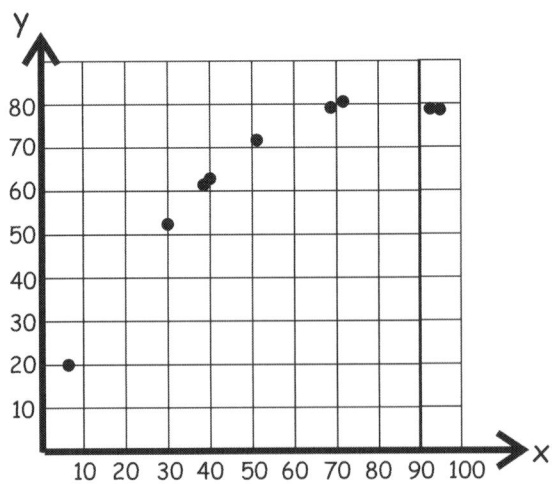

Grade 7 SOL

Vol 1 Test 1

13. Find the median for the given below data.
    21 , 17 , 18 , 22 , 18 , 14 , 17 , 14 , 17 , 18 , 15

    (A) 18   (B) 17

    (C) 14   (D) 22

14. Find the mean age from the following data.

Age	12	13	14	15	16	18	19	20	21	22	23
Frequency	1	2	2	3	2	4	3	1	1	1	1

    (A) 15.5   (B) 16.75

    (C) 16.9   (D) 17.05

15. Simplify $9x = 117$

    (A) 13   (B) -9

    (C) 10   (D) 14

Grade 7 SOL
Vol 1 Test 1

16. Translate the verbal expression to an algebraic expression.

    The 4th power of w is 9.

    (A) $w^4 = 9$ (B) $\frac{4}{2} = 9$

    (C) $w - 4 = 9$ (D) $w + 4 = 9$

17. Simplify

    $(-24) + (-22)$

    (A) -42   (B) -54

    (C) -46   (D) -30

18. Which of the following is the correct statement for the given expression?

    n^3

    (A) A number plus 3   (B) 3 cubed

    (C) The quotient of a number and 3   (D) A number cubed

19. Given below is an expression.
    What symbol shall be placed in the below box?

    $41a - 23$? 19

    (A) =   (B) +

    (C) >=   (D) None

20. Translate the verbal expression to an algebraic expression.

> The difference of z and 22 is less than 14

(A) 22 - z >= 14

(B) z - 22 < 14

(C) 22 + z < 14

(D) z + 22 ≤ 14

21. Simplify

> -23 = -17 - v

(A) -6

(B) -40

(C) $1\frac{6}{17}$

(D) 6

22. Which of the below options describes

$$22g + 7 < 2 ?$$

(A) Equation

(B) Expression

(C) Inequality

(D) Less than

23. Which of the below is the geometric series ?

(A) 1 , 5 , 15 , 25 , 120 , 145

(B) 1 , 10 , 25 , 50 , 125 , 145

(C) 1 , 5 , 25 , 120 , 155 , 175

(D) 5 , 25 , 125 , 625 , 3125

24. Cathy spent $64 to buy a big box of snack bars. If each box cost $8, how many boxes did she buy?

(A) 8

(B) 7

(C) 10

(D) 9

25. Jason spent $4.64 for a sandwich at lunch today. She now has $21.20. How much money did she had originally?

(A) $25.84

(B) $22.74

(C) $30.48

(D) $16.56

26. Write an equation showing the relation between x & y.

x	1	2	5	10
y	5	6	9	14

(A) y = 3x

(B) y = x + 4

(C) y = x - 4

(D) y = 5x

27. The table below shows Math test scores of Grade 6 student
75, 62, 70, 60, 55

The teacher missed to add a student score. If the median and mode of the data is same. What is the missing score?

(A) 62

(B) 55

(C) 75

(D) 60

28. Jack purchased a video game, originally priced at $52.50. He has a coupon for 57% discount. How much did Jack paid to the store for the game ?

    (A) $23.85   (B) $25.8

    (C) $22.58   (D) $22.90

29. Mia enlarged the size of a triangle to a base of 6 cm which originally had a base of 2 cm and a height of 4 cm. What is the height of the enlarged figure ?

    (A) 12 cm   (B) 15 cm

    (C) 14 cm   (D) 18 cm

30. The mass of an orchid seed is approximately 0.0000035 gram. Written in scientific notation, that mass is equivalent to 3.5×10^n. What is the value of n ?

    (A) -8   (B) -6

    (C) -7   (D) -5

31. Find the equivalent expression to the below:
    43.43 (6909.91 - 3153.035)

    (A) 43.43 * 6909.91 * 43.43 * 3153.035

    (B) 43.43 * 6909.91 - 43.43 * 3153.035

    (C) 43.43 * 6909.91 + 43.43 * 3153.035

    (D) (43.43 * 6909.91) / (43.43 * 3153.035)

32. Which of the below is less than 840% ?

(A) $8\dfrac{2}{5}$

(B) $\dfrac{21}{250}$

(C) 840

(D) $8\dfrac{1}{2}$

33. Rhombus ABCD and EFGH are congruent. Then what is the ratio of length $\overline{AD}$ to length $\overline{EH}$?

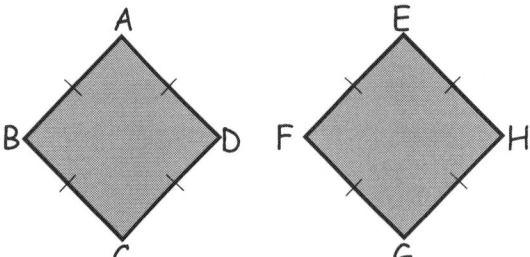

(A) 1 : 4

(B) 2 : 1

(C) 2 : 3

(D) 1 : 1

34. Polygon below is named as _____

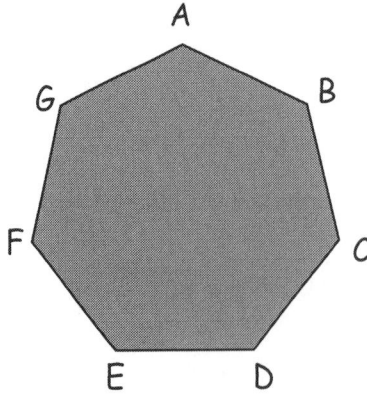

(A) Polygon

(B) Octagon

(C) Decagon

(D) Septagon

35. Find the value of x from the below triangles.

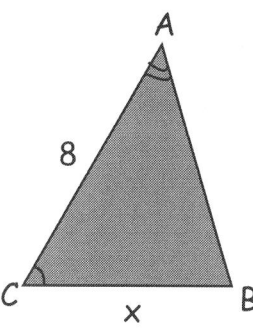
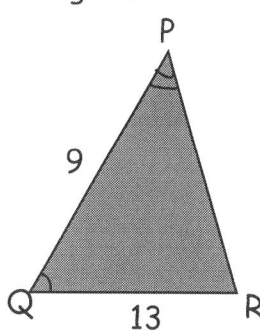

(A) $\dfrac{8}{x} = \dfrac{9}{13}$

(B) $\dfrac{x}{8} = \dfrac{9}{13}$

(C) $\dfrac{8}{x} = \dfrac{13}{9}$

(D) $8x = (13)(8)$

36. The temperature starts decreasing at a steady rate of -4° F per hour. How much will the temperature have dropped by the end of 10 hours ?

(A) -40° F

(B) -6° F

(C) 14° F

(D) 40° F

37. Below equation is an example of

55 X 101 X 37 = 37 X 55 X 101

(A) Identity Property of Multiplication

(B) Associative Property of Addition

(C) Distributive Property

(D) Commutative Property of Multiplication

Grade 7 SOL

Vol 1 Test 1

38. Which of the below is the property of a trapezoid ?

    (A) Two pair of opposite sides is parallel
    (B) Diagonals intersect each other in the unequal ratio
    (C) Two adjacent angles are supplementary
    (D) Opposite sides are of equal length

39. Which of the below is the property of a parallelogram ?

    (A) Opposite sides are parallel and equal to each other
    (B) Opposite angles are not equal
    (C) Diagonals don't bisect each other
    (D) Opposite sides are not equal in length

40. Which of the following polygons has two times as many angles as a square ?

    (A) Hexagon   (B) Decagon

    (C) Square   (D) Octagon

41. Four identical poker chips - one red chip, one blue chip, one yellow chip, and one white chip - are placed in a paper bag. If you randomly draw two of the four chips out of the paper bag, what is the probability that you will draw the red and blue chips ?

    (A) $\frac{1}{8}$ (B) $\frac{1}{6}$
    (C) $\frac{1}{4}$ (D) $\frac{1}{3}$

42. Express the given fraction as a percentage.

$$\frac{15}{16}$$

(A) 0.9375% (B) 953.75%

(C) 9375% (D) 93.75%

43. A local ice cream store donates 2% of its sales of every Sunday to a local charity. Based on the information given

   (A) Total sales amount of last Friday was $1200 and donated $24
   (B) Total sales amount of last Friday was $1100 and donated $24
   (C) Total sales amount of last Friday was $120 and donated $24
   (D) Total sales amount of last Friday was $24 and donated $1200

44. The average car sales at zoom cars across various days are plotted as the box and whisker plot.

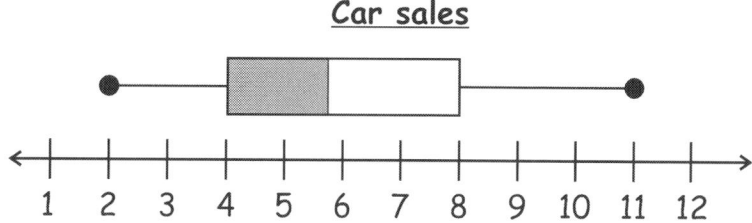

Car sales

Find the range of the test scores based on the above data

(A) 5 (B) 7

(C) 10 (D) 9

Grade 7 SOL

Vol 1 Test 1

Study the following table carefully to answer the questions 45-49.
A number of students appeared (A) and failed (F) in SOL's in the below five grades over the years.

Years	Classes									
	VI		VII		VIII		IX		X	
	A	F	A	F	A	F	A	F	A	F
2012	75	13	77	08	85	11	74	08	68	08
2013	67	17	80	09	83	06	79	06	69	15
2014	65	08	72	15	79	09	70	07	75	14
2015	69	06	66	11	77	12	71	04	84	09
2016	73	11	67	10	72	10	74	10	83	11
2017	72	12	76	07	84	05	80	05	72	05
2018	70	07	80	08	77	03	81	09	85	08

45. What is the number of passed students for all the classes together, in the Year 2018?

    (A) 305
    (B) 420
    (C) 358
    (D) 550

46. Find the average number of failed students from grade 8 for the given years.

    (A) 8
    (B) 9
    (C) 7
    (D) 11

Grade 7 SOL
Vol 1 Test 1

47. Find the ratio of the total number of passed students to total number of failed students in the year 2015.

    (A) 90:11   (B) 367:42

    (C) 42:325   (D) 35:233

48. Find the total percentage of passed students of grade 9 from all the years together. (Round to 2 decimals)

    (A) 91.88   (B) 74.92

    (C) 85.95   (D) 91.52

49. Which of the below grade has the minimum number of passed students over the years ?

    (A) Grade 8   (B) Grade 9

    (C) Grade 7   (D) Grade 6

50. David purchased a game for $13.50. Find the selling price with a 5% tax added to it

    (A) $14.18   (B) $15.52

    (C) $16.20   (D) $12.82

Grade 7 SOL

Vol 1 Test 2

1. Simplify

   $(15 - (-5)) \div (6 - 4)$

   (A) 10   (B) 6

   (C) 0   (D) 13

2. Simplify

   $(-2) - (-11)$

   (A) 9   (B) -6

   (C) -3   (D) 18

3. Express the below number in scientific notation

   74600

   (A) 74.6×10^5 (B) 74.6×10^{-5}

   (C) 7.46×10^5 (D) 7.46×10^4

4. Maria reduced the size of a photo to a width of 6 in. What is the new height if it was originally 28 in tall and 42 in wide?

   (A) 42 in   (B) 4 in

   (C) 1 in   (D) 196 in

5. Simplify

   $18 - (10 - (-9)^2)$

   (A) 78   (B) 96

   (C) 89   (D) 94

6. Which of the below graphs is an example of "Rotation 90° Counterclockwise about the origin".

(A)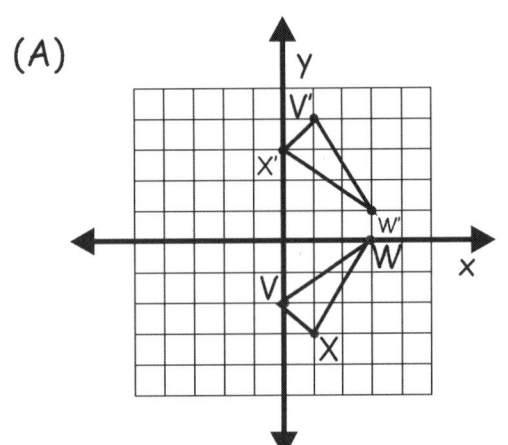

(B)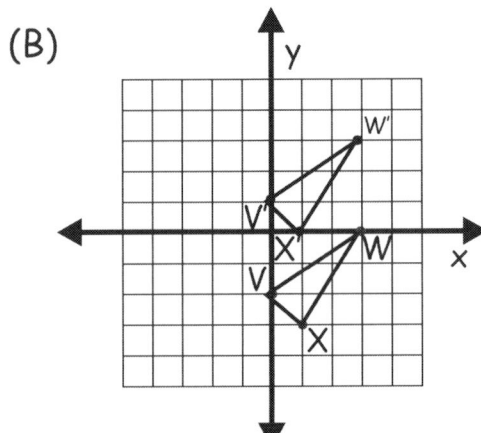

(C)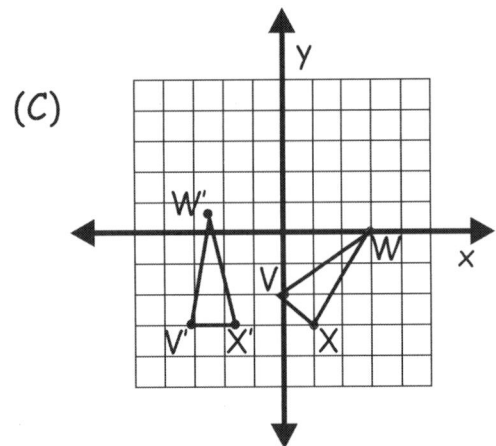

(D)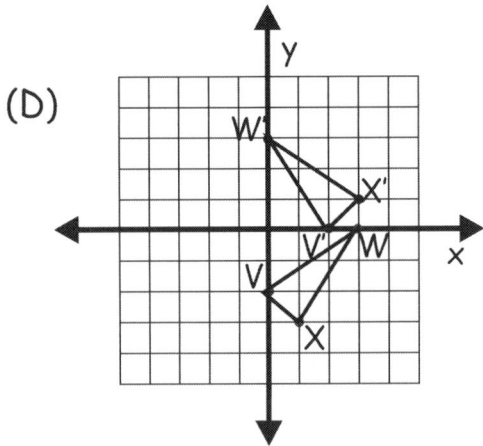

7. Determine the quadrant of the below point belongs to.
(-8, 12)

(A) Quadrant I

(B) Quadrant II

(C) Quadrant III

(D) Quadrant IV

Grade 7 SOL

Vol 1
Test 2

8. Find the volume of the cylinder with a radius of 6 m and a height of 4m.

(A) 388 m³ (B) 384.5 m³

(C) 334.5 m³ (D) 452.4 m³

9. Find the area of the enclosure.

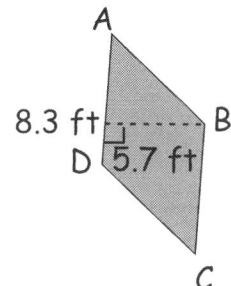

(A) 52.71 ft² (B) 23.7 ft²

(C) 47.31 ft² (D) 94.62 ft²

10. Find the total surface area of the given three dimensional shape.

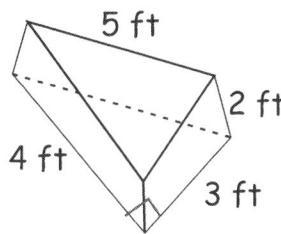

(A) 33.7 ft² (B) 27.6 ft²

(C) 30..6 ft² (D) 36 ft²

11. A class has 20 boys and 15 girls. If one representative from each sex has to be chosen, in how many ways this can be done ?

(A) 400 (B) 300

(C) 425 (D) 350

12. Which of the below scatter plots can be plotted using the given data ?

x	y
1,000	50
1,000	90
3,000	70
7,000	70
7,000	90

x	y
8,000	20
8,000	60
8,000	80
9,000	30
10,000	80

(A)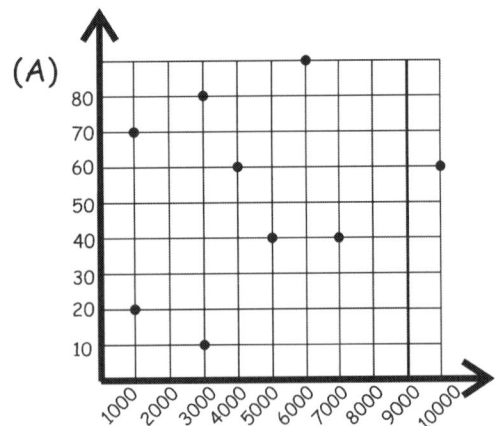

(B)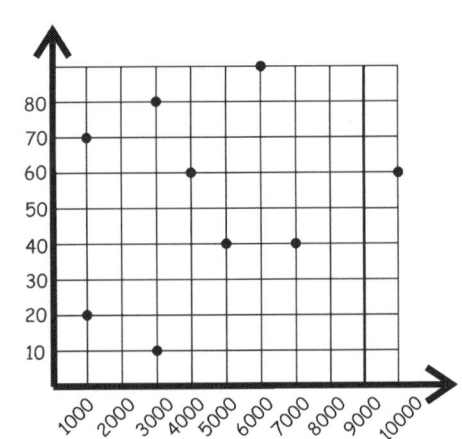

(C)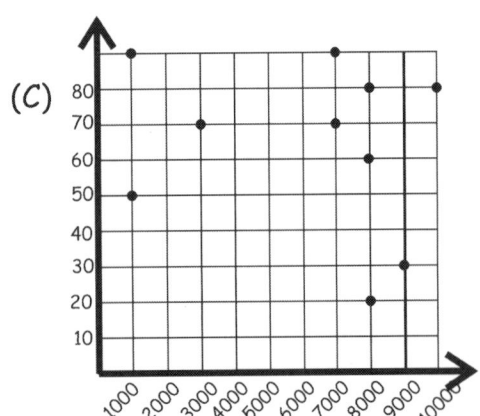

(D)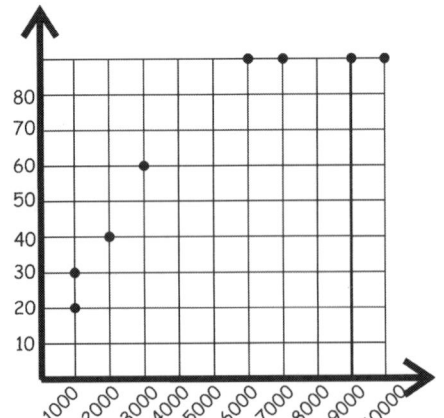

Grade 7
SOL

Vol 1
Test 2

13. Find the median for the given below data.
45 , 57 , 52 , 52 , 53 , 42 , 49 , 53 , 51 , 43

(A) 51.5 (B) 50.5

(C) 52.5 (D) 45.5

14. Find the mean score from the following data.

Score	31	42	44	45	48	49	50	51	52	53	54
Frequency	1	1	2	1	1	3	1	2	4	3	3

(A) 50 (B) 48.66

(C) 47.36 (D) 49.18

15. Simplify   -9x = 162

(A) 12 (B) 9

(C) -18 (D) 18

16. Translate the verbal expression to an algebraic expression.

   Half of n is less than or equal to 35.

   (A) $2 - n \leq 35$ (B) $n^2 \leq 35$

   (C) $\frac{n}{2} \leq 35$ (D) $n - 2 \leq 35$

17. Simplify

   $7 + (-19)$

   (A) -12 (B) -6

   (C) -29 (D) 5

18. Which of the following is the correct statement for the given expression?

   $8^x > 47$

   (A) 8 decreased by x is greater than 47
   (B) 8 to the x is greater than 47
   (C) x times 8 is greater than 47
   (D) 8 less than x is greater than 47

19. Given below is an equation.
    What symbol shall be placed in the below box?

   $5b + 13$? 32

   (A) − (B) +

   (C) >= (D) =

Grade 7 SOL

Vol 1 Test 2

20. Translate the verbal expression to an algebraic expression.

    15 less than w is 20

    (A) $15^w = 20$ (B) $w - 15 = 20$

    (C) $15 - w = 20$ (D) $w + 15 = 20$

21. Simplify

    $$-5a = -70$$

    (A) -75   (B) -12

    (C) 14   (D) -65

22. Which of the below options describes $5y - 32 = 3$?

    (A) Equation   (B) Expression

    (C) Inequation   (D) Constant

23. Which of the below is the geometric series ?

    (A) $1, \dfrac{2}{3}, \dfrac{5}{9}, \dfrac{7}{27}, \dfrac{16}{81}$

    (B) $1, \dfrac{2}{3}, \dfrac{4}{9}, \dfrac{8}{27}, \dfrac{16}{81}$

    (C) $\dfrac{1}{3}, \dfrac{2}{3}, \dfrac{5}{9}, \dfrac{8}{27}, \dfrac{16}{81}$

    (D) $1, \dfrac{1}{3}, \dfrac{45}{9}, \dfrac{7}{27}, \dfrac{16}{81}$

24. A cake recipe needs 3 cups of milk. Susan accidentally added 8 cups of milk. How many more cups did she put in?

(A) 2

(B) 1

(C) 4

(D) 5

25. Noah and his best friend got a cash prize. They split the money evenly, each getting $23.76. How much money was the cash prize?

(A) $47.52

(B) $48.17

(C) $11.88

(D) $51.80

26. Write an equation showing the relation between x & y.

x	1	2	3	6
y	3	6	9	18

(A) y = x + 2

(B) y = 2x

(C) y = 3x

(D) y = x + 3

27. The table below shows goals made by Ben in his soccer tournament.

8, 9, 7, 3, 8

The teacher missed to add a student score. If the median and mode of the data is same. What is the missing score?

(A) 8

(B) 5

(C) 10

(D) 9

28. A writing desk is priced at $249.50 and the discount offered on it is 55%, calculate the selling price.

    (A) $112.27

    (B) $137.23

    (C) $112.28

    (D) $386.73

29. Lucy is drawing a picture with a height of 2 in. The original picture had width of 2 in and a length of 4 in. What is the length of Lucy's drawing?

    (A) 2 in

    (B) 8 in

    (C) 4 in

    (D) 3 in

30. The number 1.56×10^{-2} is equivalent to

    (A) 156

    (B) 0.0156

    (C) 0.156

    (D) 0.00156

31. Find the equivalent expression to the below:
    $$10.10 \ (11.11 + 101.101)$$

    (A) 10.10 * 11.11 / 10.10 * 101.101

    (B) 10.10 * 11.11 * 10.10 * 101.101

    (C) 10.10 * 11.11 - 10.10 * 101.101

    (D) 10.10 * 11.11 + 10.10 * 101.101

Grade 7 SOL

Vol 1 Test 2

32. Which of the below is greater than 560% ?

   (A) 560

   (B) $5\frac{3}{5}$

   (C) $\frac{7}{125}$

   (D) $\frac{14}{25}$

33. Parallelogram ABCD and LMNO are congruent. Then what is the ratio of length $\overline{CD}$ to length $\overline{GH}$?

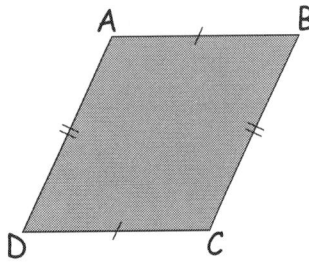
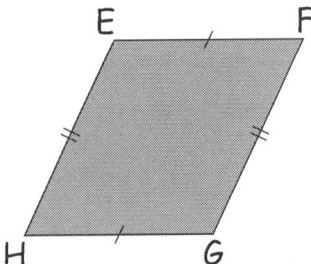

   (A) 1 : 1

   (B) 2 : 1

   (C) 2 : 3

   (D) 1 : 4

34. Polygon below is named as _____

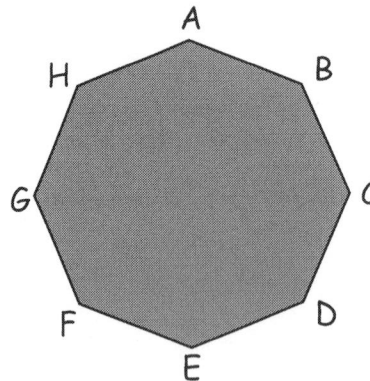

   (A) Octagon

   (B) Decagon

   (C) Polygon

   (D) Septagon

Grade 7
SOL

Vol 1
Test 2

35. Find the value of p from the below triangles.

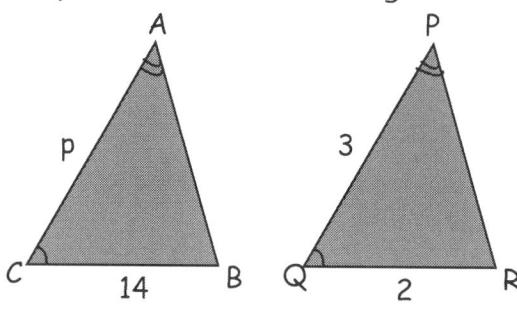

(A) $\dfrac{p}{14} = \dfrac{3}{2}$

(B) $\dfrac{p}{14} = \dfrac{5}{3}$

(C) $\dfrac{14}{p} = \dfrac{3}{2}$

(D) $14p = 6$

36. Tracy takes $2500 debt every year for four years to pay his college fees. How much is Tracy's debt at the end of 4 years ? (Assuming no interest on the loan amount)

(A) 5000

(B) -10000

(C) -2500

(D) 10000

37. Below equation is an example of
$$71 * 1 = 1 * 71$$

(A) Identity Property of Multiplication
(B) Associative Property of Addition
(C) Distributive Property
(D) Commutative Property of Multiplication

38. Which of the below is the property of a rhombus ?

(A) Every rhombus is a not parallelogram
(B) Diagonals are not perpendicular bisectors of each other
(C) Every rhombus is a kite with un equal sides of length
(D) All sides are congruent

Grade 7 SOL   **Vol 1 Test 2**

39. Which of the below is the property of a kite?

    (A) One of the diagonals is the perpendicular bisector of another
    (B) Diagonals of a kite intersect at acute angles
    (C) Two distinct pairs of adjacent sides are congruent
    (D) Angles between unequal sides are unequal.

40. Which of the following polygons has two more angles than a septagon?

    (A) Hexagon   (B) Nonagon

    (C) Septagon   (D) Octagon

41. If you roll three normal, six-sided dice in a completely random manner, what is the probability that the three numbers that come up on the dice will all be identical?

    (A) $\dfrac{1}{36}$ (B) $\dfrac{1}{12}$

    (C) $\dfrac{1}{18}$ (D) $\dfrac{1}{216}$

42. Express the given fraction as a percentage.

    $$\dfrac{4}{5}$$

    (A) 8000%   (B) 0.8%

    (C) 150%   (D) 80%

43. A local smoothie shop donates 5% of its sales of every Friday to a local charity. Based on the information given

    (A) Total sales amount of last Friday was $37 and donated $750
    (B) Total sales amount of last Friday was $100 and donated $95
    (C) Total sales amount of last Friday was $750 and donated $37.50
    (D) Total sales amount of last Friday was $250 and donated $125

A robot manufacturing company spends $2880k every year as given in the pie chart below. Answer the questions 44 - 49.

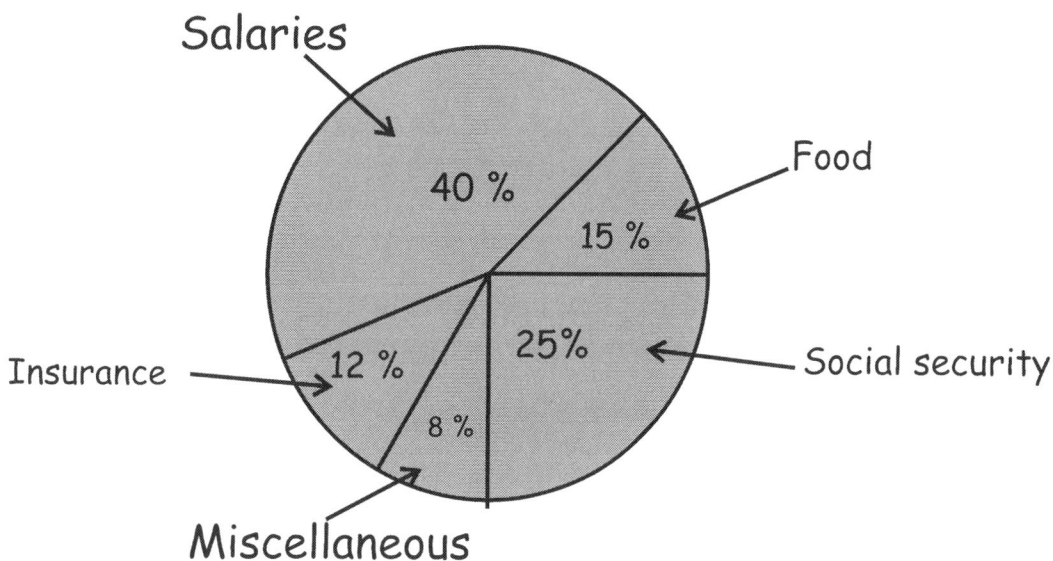

44. Find the amount was spent towards salaries ?

    (A) $1152k (B) $1252k

    (C) $1100k (D) $1052k

45. How much money is saved if the budget allocated to miscellaneous is not spent and food went 10% over the budget ?

    (A) $1972k (B) $187.2k

    (C) $2736k (D) $2076k

Grade 7
SOL

Vol 1
Test 2

46. How much money was spent on food ?

   (A) 432k (B) 335k

   (C) 444k (D) 533k

47. How much money was spent on insurance ?

   (A) 349k (B) 395.9k

   (C) 312.6k (D) 345.6k

48. How much more money was spent on social security than insurance ?

   (A) 377.38k (B) 384.49k

   (C) 374.4k (D) 399.99k

49. Find the ratio between the amount spent on food and social security ?

   (A) 2:5 (B) 5:3

   (C) 3:5 (D) 2:3

50. The average test scores of grade 8 students are plotted as the box and whisker plot

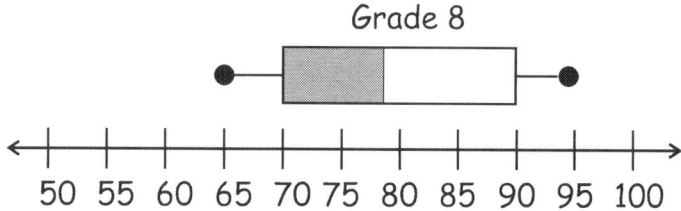

Find the range of the test scores based on the above data

(A) 20

(B) 45

(C) 30

(D) 35

Grade 7 SOL

Vol 1 Test 3

1. Simplify

    (5 - 11) · 11 ÷ (-2)

    (A) 78   (B) 92

    (C) 93   (D) 83

2. Simplify

    (-15) - 15

    (A) -44   (B) -30

    (C) -23   (D) -16

3. Express the given number in scientific notation.

    0.0000085

    (A) 0.85×10^{-6} (B) 85×10^{6}

    (C) 8.5×10^{6} (D) 8.5×10^{-6}

4. If you can buy five seedless watermelons for $9, then how many can you buy with $18 ?

    (A) 32   (B) 9

    (C) 10   (D) 8

5. Simplify

    15(17 - ((-21) ÷ (-3)))

    (A) 150   (B) 148

    (C) 162   (D) 156

6. Which of the below is an example of "Rotation 90° counterclockwise about the origin"?

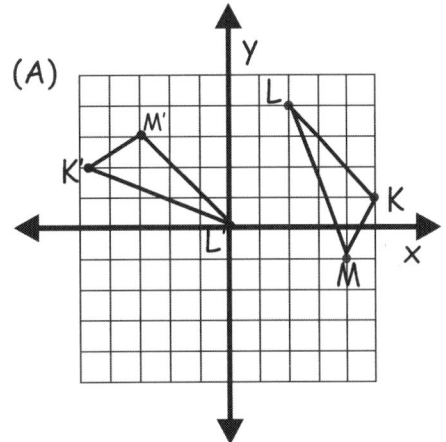

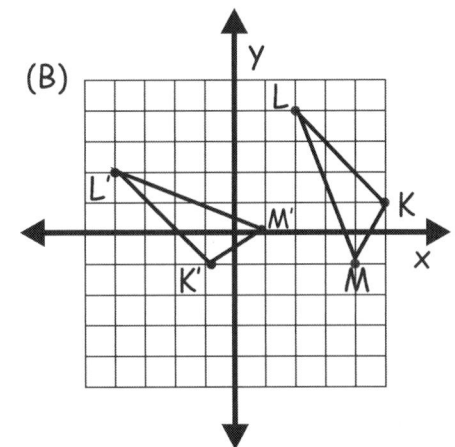

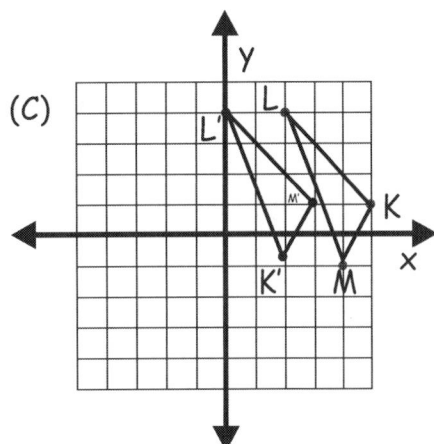

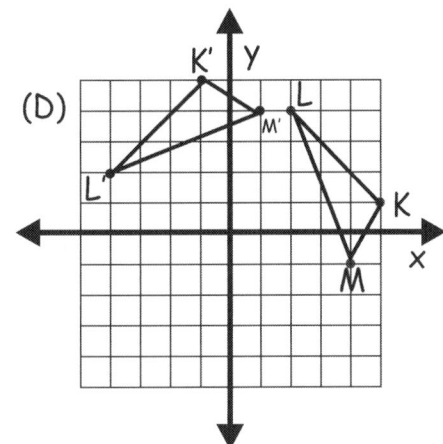

7. Determine the quadrant of the below point belongs to.

(-33, -33)

(A) Quadrant I

(B) Quadrant II

(C) Quadrant III

(D) Quadrant IV

8. Find the volume of the cylinder with a radius of 5 mi and a height of 4 mi.

(A) 70.7 mi³

(B) 314.2 mi³

(C) 78.6 mi³

(D) 157.1 mi³

9. Find the area of the enclosure.

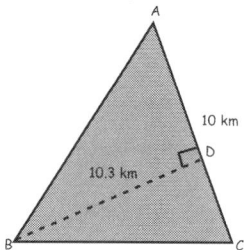

(A) 51.5 km²

(B) 25.8 km²

(C) 103 km²

(D) 53.5 km²

10. Find the total surface area of the given three dimensional shape.

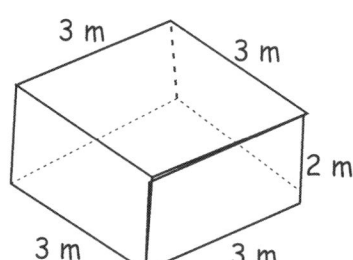

(A) 42 m²

(B) 34 m²

(C) 24 m²

(D) 39 m²

11. How many different outcomes arise from first tossing a coin and then rolling a die ?

(A) 12

(B) 10

(C) 8

(D) 14

12. Which of the below scatter plots can be plotted using the given data ?

X	y
1	1
3	4
4	8
5	6
6	30

X	y
8	25
8	30
9	20
10	15
10	11

(A)
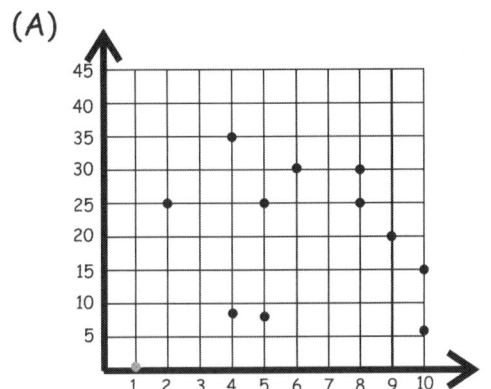

(B)
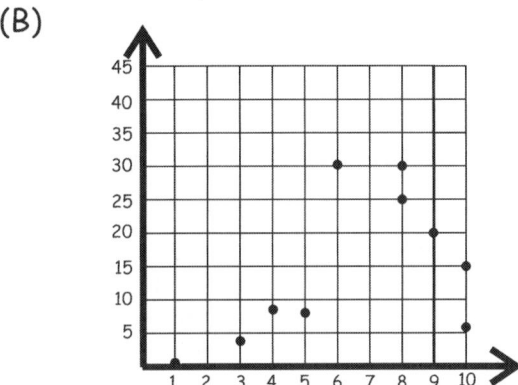

(C)
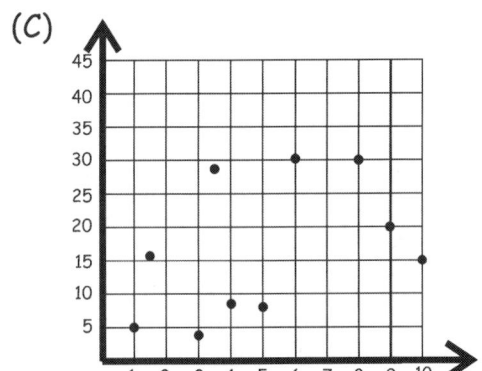

(D)
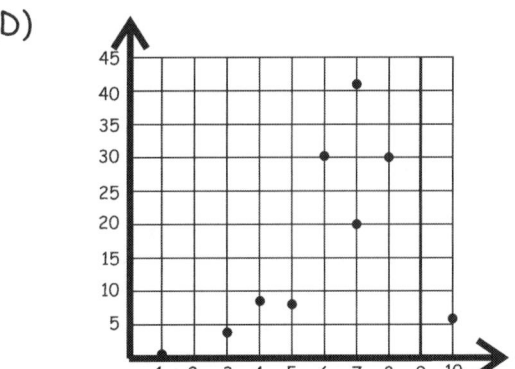

Grade 7 SOL   **Vol 1 Test 3**

13. Find the median for the given below data.
    25.1 , 33 , 41.3 , 21.9 , 31.7 , 39 , 35.2 , 21.9 , 27 , 28.9 , 25.9

    (A) 25.1   (B) 35.2

    (C) 27   (D) 28.9

14. Find the mean of the following data.

Hours	Frequency
5.75	1
6.5	1
6.75	7
7	2
7.25	1
7.5	4
7.75	2
8	1
8.25	2
8.5	2

    (A) 7.29   (B) 6.94

    (C) 6.72   (D) 7.48

15. Simplify

    $$-7 = \frac{a}{14}$$

    (A) -21   (B) 7

    (C) -98   (D) $-\frac{1}{2}$

16. Translate the verbal expression to an algebraic expression.

> n squared is greater than or equal to 20.

(A) $2 - n \geq 20$

(B) $n^2 \geq 20$

(C) $2 + n > 20$

(D) $2^2 \geq 20$

17. Simplify

> $(-5) - (-10)$

(A) 5

(B) 13

(C) 6

(D) –2

18. Which of the following is the correct statement for the given expression?

> $x - 12 > 8$

(A) 12 divided by x is greater than 8
(B) x cubed is greater than 8
(C) x squared is greater than 8
(D) x decreased by 12 is greater than 8

19. Given below is an inequality.
What symbol shall be placed in the below box?

> $11x - 2$? 9

(A) =

(B) +

(C) >=

(D) None

Grade 7 SOL
Vol 1 Test 3

20. Translate the verbal expression to an algebraic expression.

   | The quotient of a number and 3 is equal to 15 |

   (A) $\dfrac{3}{n}$

   (B) n - 3 = 15

   (C) $\dfrac{n}{3}$ = 15

   (D) 3^n = 15

21. Simplify

   $$a + (-15) = -12$$

   (A) -27

   (B) 3

   (C) -9

   (D) $\dfrac{4}{5}$

22. Which of the below options describes 99x - 1 ?

   (A) Expression

   (B) Greater than

   (C) Inequation

   (D) Equation

23. Which of the below is the geometric series ?

   (A) 1, 4, 8, 24, 36, 108

   (B) 1, 6, 12, 24, 108, 324

   (C) 1, 4, 12, 36, 124, 248

   (D) 4, 12, 36, 108, 324

Grade 7 SOL

Vol 1 Test 3

24. Charlie and seven of his friends went for dinner to a restaurant. They split evenly. Each person paid $10. What was the total bill ?

   (A) $70   (B) $86

   (C) $80   (D) $1.43

25. Last week Leo ran 8 miles less than Nathan. Leo ran 12 miles this week. How many miles did Nathan ran ?

   (A) 18   (B) 20

   (C) 4   (D) 28

26. Write an equation showing the relation between x & y.

x	6	8	10	16
y	30	28	16	20

   (A) y = 6 + a   (B) y = 5a

   (C) y = x + 30   (D) y = 36 - a

27. The table below shows number of brand A shirts sold across various stores.

   259 , 234 , 252 , 266 , 267 , 269 , 259 , 237

   If the median and mode of the data is not altered after adding the missing sales, what could be the missing sales value ?

   (A) 268   (B) 235

   (C) 266   (D) 259

Grade 7 SOL

Vol 1 Test 3

28. David has a 5% discount coupon for any purchases at store A. All hand bags are priced at $99.99. What is the selling price of one bag?

(A) $79.99

(B) $89.50

(C) $94.99

(D) $104.99

29. A box of cherries cost $2. How many boxes of cherries can Tom buy for $20?

(A) 10

(B) 40

(C) 11

(D) 9

30. The size of a certain type of molecule is 0.00009078 inch. If this number is expressed as 9.078×10^n, what is the value of n?

(A) -5

(B) -8

(C) 5

(D) 8

31. Find the equivalent expression to the below:

    72.18 (29.013 + 909.19)

(A) 72.18 * 29.113 + 72.18 * 999.19

(B) 72.18 * 29.013 - 72.18 * 909.19

(C) 72.18 * 29.013 * 72.18 * 909.19

(D) 72.18 * 29.013 + 72.18 * 909.19

32. Which of the below is greater than 325% ?

   (A) $3\dfrac{29}{100}$

   (B) $\dfrac{4}{13}$

   (C) $3\dfrac{1}{4}$

   (D) 325

33. Circle 'A' and Circle 'B' are congruent and their radii are congruent by a scale factor of 4. Then what is the ratio of their areas ?

   (A) 1 : 4

   (B) 1 : 16

   (C) 2 : 3

   (D) 1 : 2

34. Polygon below is named as _____

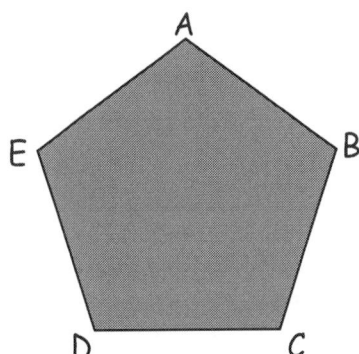

   (A) Quadrilateral

   (B) Polygon

   (C) Pentagon

   (D) Septaon

35. Find the value of n from the below triangles.

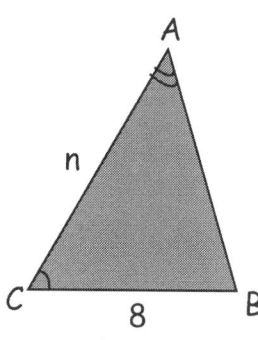 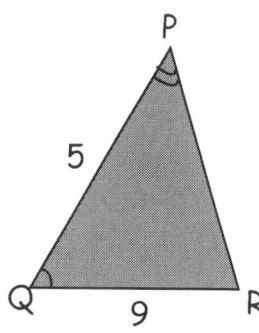

(A) 8n = 72

(B) $\dfrac{n}{8} = \dfrac{5}{9}$

(C) $\dfrac{8}{n} = \dfrac{5}{9}$

(D) $\dfrac{n}{8} = \dfrac{9}{5}$

36. A submarine is 1050 meters below the sea level. Later it ascended 300 meters, But then descended another 275 meters and finally stopped at this location. Where was the submarine in relation to sea level when it stopped ?

(A) 1625

(B) -1625

(C) 1025

(D) -1025

37. Below equation is an example of
$31 * (21 + 51) = 31 * 21 + 31 * 51$

(A) Identity Property of Multiplication

(B) Associative Property of Addition

(C) Distributive Property

(D) Commutative Property of Multiplication

Grade 7 SOL

Vol 1 Test 3

38. Which of the below is the property of a square ?

    (A) Each interior angle measures 90°
    (B) All sides are not congruent
    (C) Diagonals are not perpendicular bisectors of each other
    (D) Every square is a not a rectangle and not a rhombus

39. Two six sided dice are rolled.
    What is the probability of getting a score greater than 3 ?

    (A) $\dfrac{33}{36}$ (B) $\dfrac{31}{36}$

    (C) $\dfrac{1}{6}$ (D) $\dfrac{5}{6}$

40. Which of the following polygons has four less angles than a decagon ?

    (A) Hexagon   (B) Pentagon

    (C) Septagon   (D) Octagon

41. A certain game spinner has eight equal-sized sections numbered 1, 2, 3, 2, 4, 3, 2 and 5. If you spin this spinner twice (randomly - no cheating!), what is the probability that you will get a two both times ?

    (A) $\dfrac{3}{8}$ (B) $\dfrac{6}{8}$

    (C) $\dfrac{9}{64}$ (D) $\dfrac{9}{16}$

42. Express the given fraction as a percentage.

$$\frac{11}{50}$$

(A) 0.22% (B) 2.2%

(C) 11.5% (D) 22%

43. A local bakery donates 10% of its sales of every Saturday to a local charity. Based on the information given

(A) Total sales amount of last Friday was $950 and donated $100
(B) Total sales amount of last Friday was $165 and donated $1650
(C) Total sales amount of last Friday was $770 and donated $70
(D) Total sales amount of last Friday was $1650 and donated $165

44. The cost of a chair is $249.50 and a 5% of tax is added to it. Calculate the selling price

(A) $237.55 (B) $274.25

(C) $261.98 (D) $12.48

Grade 7
SOL

Vol 1
Test 3

The annual expenditure of an average house hold in the Washington DC is given in the below pie chart. Answer the questions 45 - 50.

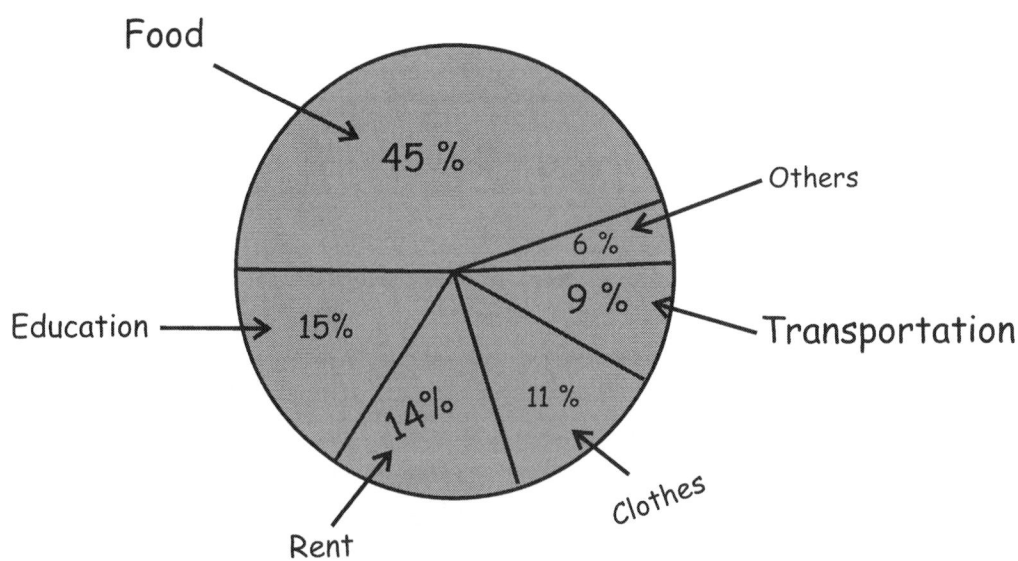

45. Find the ratio of expenses made on education and food ?

   (A) 3:1   (B) 1:3

   (C) 2:5   (D) 2:3

46. Find the ratio of expenses made on rent and others ?

   (A) 7:3   (B) 3:7

   (C) 5:7   (D) 7:2

Grade 7
SOL

Vol 1
Test 3

47. Find the ratio of expenses made on clothes and transportation ?

   (A) 1:11   (B) 6:11

   (C) 11:9   (D) 9:11

48. Tony's annual income is $25,000. Find the amount spent on rent and food ?

   (A) $17,250   (B) $16,750

   (C) $11,250   (D) $14,750

49. Tony's annual income is $25,000. Find the amount spent on rent and clothes ?

   (A) $6,250   (B) $6,575

   (C) $6,650   (D) $7,625

50. Tony's annual income is $25,000. Find the amount spent on education ?

   (A) $3,350   (B) $3,880

   (C) $3,745   (D) $3,750

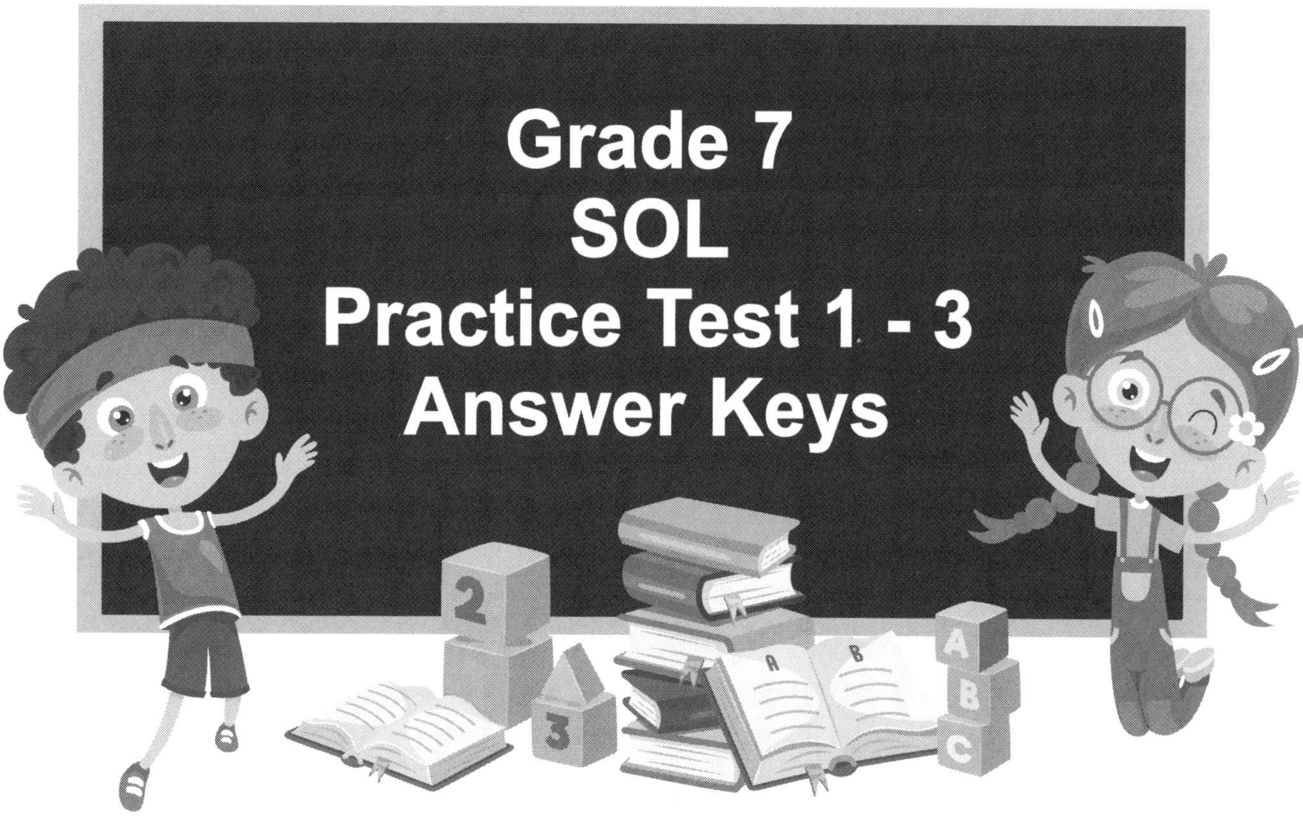

Grade 7 SOL Practice Test 1 - 3 Answer Keys

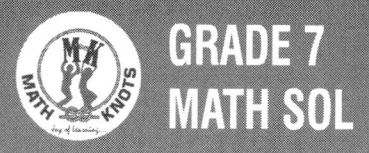

Answer Key Test - 1

1. C
2. B
3. C
4. D
5. A
6. C
7. A
8. D
9. D
10. B
11. A
12. C
13. B
14. D
15. A

GRADE 7 MATH SOL

TEST - 1 KEYS

16. A
17. C
18. D
19. B
20. B
21. D
22. C
23. D
24. A
25. A
26. B
27. A
28. C
29. A
30. B

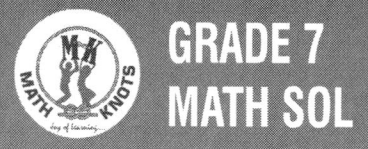

GRADE 7 MATH SOL

TEST - 1 KEYS

31. B
32. A
33. D
34. D
35. A
36. A
37. D
38. C
39. A
40. D
41. B
42. D
43. A
44. D
45. C
46. A

47. B

48. D

49. D

50. A

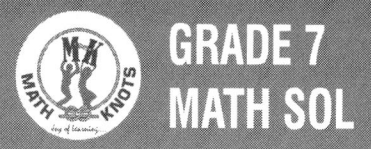

Answer Key Test - 2

1. A
2. A
3. D
4. B
5. C
6. D
7. B
8. D
9. C
10. D
11. B
12. C
13. A
14. D
15. C

GRADE 7 MATH SOL

TEST - 2 KEYS

16. C

17. A

18. B

19. D

20. B

21. C

22. A

23. B

24. D

25. A

26. C

27. A

28. C

29. C

30. B

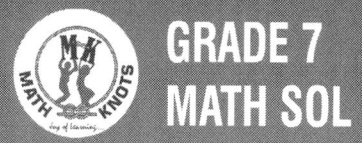

GRADE 7 MATH SOL

TEST - 2 KEYS

31. D
32. B
33. A
34. A
35. A
36. B
37. A
38. D
39. C
40. B
41. D
42. D
43. C
44. A
45. B
46. A

GRADE 7 MATH SOL

TEST - 2 KEYS

47. D

48. C

49. C

50. C

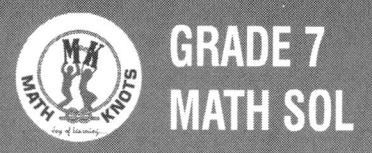

Answer Key Test - 3

1. A
2. B
3. D
4. C
5. A
6. D
7. C
8. B
9. A
10. A
11. A
12. B
13. D
14. A
15. C

GRADE 7 MATH SOL

TEST - 3 KEYS

16. B
17. A
18. D
19. C
20. C
21. B
22. A
23. D
24. C
25. B
26. D
27. D
28. C
29. A
30. A

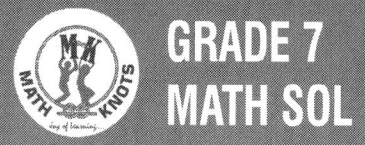

GRADE 7 MATH SOL

TEST - 3 KEYS

31. D
32. C
33. B
34. C
35. B
36. D
37. C
38. A
39. A
40. A
41. C
42. D
43. D
44. C
45. B
46. A

47. C

48. D

49. A

50. D

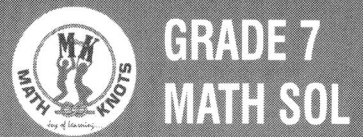
GRADE 7 MATH SOL

SCORE

GRADE 7 MATH SOL

SCORE

Score calculation

If you get this many times correct :	Then your converted scale scor is :
0	000
1	217
2	249
3	268
4	282
5	293
6	303
7	311
8	319
9	325
10	331
11	337
12	343
13	348
14	353
15	357
16	362
17	366
18	371
19	375

GRADE 7 MATH SOL

20	379
21	383
22	387
23	391
24	395
25	399
26	403
27	407
28	411
29	415
30	419
31	423
32	427
33	432
34	436
35	441
36	445
37	450
38	456
39	461

40	467
41	473
42	480
43	487
44	496
45	505
46	516
47	531
48	550
49	582
50	600

Made in the USA
Columbia, SC
12 March 2025